Matchmaking

Grateful acknowledgment is made to:
Animals Animals: pages 11 (top – Ray Richardson), 14
        (H. Ansloos), 26 (Ray Richardson), 28 (Ray Richardson)
Attilio Calegari/Overseas: pages 6, 25 (bottom)
Anthony Bannister: page 18
George Bernard: page 21 (top) and front cover
M. F. Black: page 23 (bottom)
Raymond Blythe: page 24
Chris Catton: page 15 (bottom)
David Cayless: page 15 (top)
Martyn Chillmaid: page 12
J. A. L. Cooke: page 9 (bottom)
Roy Coombes: page 25 (top)
Michael Fogden: pages 4, 5, 16 (bottom), 17
R. J. B. Goodale: page 29
David Macdonald: page 19 (top right and left)
Mantis Wildlife Films: pages 22, 27 and title page
T. S. McCann: pages 13 (bottom), 21 (bottom)
Sean Morris: page 23 (top)
Stan Osolinski: page 10
Avril Ramage: pages 13 (top), 16, (top)
D. J. Saunders: page 8
P. K. Sharpe: page 19 (bottom)
Alastair Shay: page 20
D. H. Thompson: page 9 (top)
Maurice Tibbles: page 7 (bottom) and back cover
P. and W. Ward: page 11 (bottom)
Special thanks to Bruce Coleman Ltd., for the Lesser Bird of
        Paradise, page 7 (top)

First American edition, 1987
Copyright © 1987 by Oxford Scientific Films.
Originated and published in Great Britain by
André Deutsch Limited, 1987.
Printed and bound by Proost, Turnhout, Belgium
Library of Congress Cataloging-in-Publication Data
Matchmaking.
   Summary: Text and photographs depict the courtship
and sexual behavior of animals including the golden
toad, elephant seal, garden spider, and bower bird.
   1. Sexual behavior in animals – Juvenile literature.
[1. Sexual behavior in animals. 2. Animals – Habits
and behavior] I. Oxford Scientific Films.
QL761.M365   1987   591.56′2   87–2538
ISBN 0–399–21451–8
G.P. Putnam's Sons, 51 Madison Avenue
New York, New York 10010
First Impression

Banded Iguanas courting

Matchmaking

Oxford Scientific Films
edited by
CHRIS CATTON

G.P. Putnam's Sons, New York

A NEW LIFE BEGINS

Golden Toads. In the jungles of Central America, new life is about to begin. First the male golden toads gather around a pool of water.

When a male sees a female, he climbs onto her back and is carried into the pool. Several other males have the same idea.

Beneath the water, the female toad squeezes out a long string of large eggs. Before these can grow into tadpoles, they must be fertilized by sperm from the male.

DRESSING UP

Female animals choose their mates carefully, so the males try hard to impress them. Some make a spectacular display with bright colors.

Golden Pheasant.
When displaying to females, this male will fan his bright collar forward.

Lesser Bird of Paradise.
The male has extraordinary and beautiful feathers which he shows off to the females.

Frigate Bird.
The male displays by filling his scarlet throat pouch like a balloon. The female on his left has only a white throat.

GIVING PRESENTS

Some male animals must give the female a present before she will mate with them.

Wren. The cock wren offers hens a nest as a present. If a hen chooses to raise her young in it, she becomes his mate.

Stickleback. The male stickleback builds a nest of weed and mud. There is a tunnel through the middle of the nest in which the female will lay her eggs for the male to fertilize.

Roadrunner. The male has caught a lizard and during mating the female will reach up and grab it.

CARRYING WEAPONS

Male animals sometimes carry weapons that will help them win fights for females.

Deer.
Deer use antlers.

Bighorn Sheep. Sheep use horns.

Fiddler Crab.
Fiddler crabs
use a big
strong claw.

FIGHTING FOR A MATE

The instinct to raise a family is so powerful that male animals will sometimes fight to win their mates. Usually these fights are tests of strength, and the weaker male gives up and runs away before either is hurt. But not always.

Mute Swans. Each of these mute swans is trying to give the other a ducking by forcing his head underwater.

Stag Beetles. Male stag beetles fight by gripping their rivals between giant pincers. Can you think how stag beetles got their name?

Elephant Seals. Elephant seals rear up on their tails and then crash down on each other, biting as they fall.

Stag Beetles

Elephant Seals

WOOING WITH SOUNDS

Sounds can travel great distances, and a male who woos his mate with song can be heard even when he cannot be seen.

Red Deer. Red deer stags roar day and night throughout the breeding season.

Cockerel.
This cockerel's ancestors once lived in the jungles of Southeast Asia, where his crowing would draw hens that were ready to mate.

Eider.
With his head thrown back and chest thrust out, this eider drake courts a female with a gentle cooing.

MANY KINDS OF CROAKING

These male frogs all have different sounding croaks. Females listen to the croaking and hop toward males of their own kind.

African Bullfrog. This fat frog has a very deep croak.

Puddle Frog. The puddle frog's croak sounds rather like a sheep baaing.

Painted Reed Frog. There are many different kinds of reed frog. They all look very much alike, but are easy to tell apart if you hear them.

Red and Blue Arrow-poison Frog. This frog attracts females by croaking at night. His bright colors have nothing to do with courtship; instead they warn enemies of his poisonous skin.

SNIFFING OUT A MATE

Smells are very important in the animal world. Male animals can use perfumes to leave messages for passing females, or they can search out females by following a scent trail.

Klipspringer.
This male klipspringer is smearing perfume onto a branch.
The smell warns other males to keep away.

Capybara.
The male capybara is claiming this female as his own by rubbing his scent on her.

The scent is produced in a special gland which makes a bump on the male's nose.

Dogs. Dogs learn a lot about each other by sniffing. If a female is ready to mate, her smell attracts males from all around.

COURTING DANCES

Movements can carry messages from males to females just as well as songs or smells. A peculiar flight or jerky dance by the male will help the female to recognize him as he approaches.

Postman Butterfly. The hovering dance of the male postman butterfly begins his courtship of the female.

Scorpions.
Scorpions have poisonous stings which they use to kill their prey. One mistake in this dance could be fatal.

Wandering Albatross. The bill of an albatross is a sharp and powerful weapon. By pointing it to the sky as he dances around, the male makes it clear that he means no harm.

MINIATURE MALES

Female animals are sometimes bigger than their mates.

Orchid Mantis. The male orchid mantis is only half as long as his mate.

Panamanian Orb-weaving Spider. Can you see the little male on the belly of this fat female spider?

Garden Spider. The female garden spider is much bigger than the male. If she attacks, the male has little chance of escaping. His dance tells her that he is a male and has come to mate.

COURTING WITHOUT GETTING CAUGHT

Animals that dress in bright colors cannot blend into the background and are quickly spotted by their enemies.

Glowworm. This female glowworm has a clever trick for attracting a mate. She flashes a bright light at the end of her body. When the light is off, she is hidden to any creature that might want to eat her up.

Mallard.

The mallard drake looks splendid in his courtship finery.

But the female is dull and can hide among the reeds without being seen.

WHAT IS IT?

Both of these pictures have something to do with courtship. But what? Can you see what animal this is? And can you guess what sort of creature made this strange collection? Turn the page to find out the answers.

ANSWERS

Perhaps you could see that the creature on page 26 was a bird. If you guessed that the collection on page 27 was made by another bird you did very well indeed!

Sage Grouse. The male sage grouse hardly looks like a bird at all when he is courting, even though the females are quite ordinary.

Great Bower Bird. The great bower bird collects together sticks, stones, bottle tops, bits of glass – anything, just to impress his mate.

MORE FACTS

A NEW LIFE BEGINS (pages 4–5)

Golden Toads. Golden toads are very rare, found only in a couple of square miles of Costa Rican rain forest. Most of the year they are secretive. But when the first spring rains fall, forming breeding pools, the males congregate around them. They make only a quiet trill – it is their color that attracts the females.

DRESSING UP (pages 6–7)

Golden Pheasant. Golden pheasants live in the mountains of central China. Like other pheasants, the males have very elaborate and gaudy feathers to attract as many females as possible.

Lesser Bird of Paradise. There are forty-three species of birds of paradise, so-called because of the males' spectacular plumage used in courtship. The females' coloring is dull to camouflage them.

Frigate Bird. The five frigate bird species are large soaring seabirds which breed in the Atlantic. All perform bizarre displays which include flapping wings and rattling bills against the inflated red pouch.

GIVING PRESENTS (pages 8–9)

Wren. The cock (male) wren often builds seven or eight nests and shows them all to the hen (female) birds who visit him. Most of the nests are never used, and are called "cock nests." The hen always picks a nest that has been well made and well hidden.

When he has found a female for one nest and mated with her, the cock wren does not give up his search. He might be able to persuade a second hen to mate and occupy another of his nests. The hens will do all the incubating, but the cock wren helps feed the young. If he is the father of two broods, he will be very busy.

Stickleback. The male stickleback begins making his nest in the spring, sticking together bits of weed and mud with a special glue made in his kidneys. When the female has laid her eggs, he chases her away, fertilizes the eggs and then looks after them himself. Inside the eggs, tiny fish are growing. The male fans water over the eggs with his fins, which helps the baby fish inside to breath, and he scares away other male sticklebacks that might try to steal the eggs.

Roadrunner. Many male birds encourage females to mate with them by offering presents of food. The female needs to eat much more during the breeding season because the eggs are growing inside her.

CARRYING WEAPONS (pages 10–11)

Deer. Antlers are fearsome weapons. Each year they fall off completely in the winter and begin to grow again in the spring. At first they are covered in a soft velvet skin that feeds the bone growing beneath. Then in late summer, the skin dries up and falls away, or is rubbed off by thrashing the antlers against small bushes.

Bighorn Sheep. These sheep use their massive, curling horns to batter each other. The rams charge head to head and crash at top speed. Eventually one gets stunned by the blows and staggers away. Only rams that are evenly matched get involved in fights. When two rams meet during the breeding season they first size each other up carefully. If one sees that he is clearly going to lose the fight, he runs off quickly.

Fiddler Crab. This fiddler crab is found along the coasts of East Africa. He lives in the mud, and defends an area around his hole by waving his huge claw at all the other crabs. If another male crab comes too close, he drives it away, using his claw as a sledgehammer to batter the intruder.

The claw-waving acts as a threat to other males, but it is also a way of courting females. If a female is carrying eggs, she is not frightened by the male. Instead she walks straight toward him, and he takes her down into his burrow, where they mate.

FIGHTING FOR A MATE (pages 12–13)

Mute Swans. Male swans are called cobs, and females are pens. A pair of mute swans keeps other birds away from the area around their nest all through the breeding season. This is their territory, and any cob that strays inside will be attacked. When the intruder has been driven away, cob and pen swim toward each other and press their breasts and necks together, reassuring themselves and strengthening their bond. Mute swans are not silent at all, but may have gotten their name because they call to each other more softly than other species of swans.

Stag Beetles. Only the male stag beetles have these vicious looking pincers, which look like antlers – hence the beetle's name. The pincers are used only for fighting, and females manage perfectly well without them. The beetles try to grab their opponents around the middle and lift them from the ground. The loser cannot fight back in this position, but his hard body saves him from being squashed. Often the winner carries the loser to the edge of a branch and throws him off.

Elephant Seals. Elephant seals weigh only about half as much as real elephants, but each bull is still heavier than a big car. At breeding time, hundreds of females lie on the beaches of their Antarctic home, and a single male may be able to keep fifty females to himself by fighting off lesser bulls. Their fights are fierce, and old bulls are often covered in deep scars.

WOOING WITH SOUNDS (pages 14–15)

Red Deer. During the breeding season, red deer stags roar continuously. In the misty hills and valleys where they live, their sound can be heard by does that are far out of sight. The loud roaring is hard work, and rival stags have roaring contests to test each other's strength – a much less dangerous alternative to fighting.

Cockerel. Birds that live in woodlands cannot be seen very far through the trees, so instead of dancing or showing off their brightly colored feathers, they sing to each other. In his jungle home, each cockerel controls a patch of ground, and any intruding males are quickly chased away. His crowing draws the hens to mate with him. In a farmyard, although the hens are all around, the cockerel still crows.

Eider. Most drakes use color or movement to impress their females, but the cooing display is an important part of the eider's courtship. Perhaps this is because they court at sea, and in rough weather they often lose sight of each other behind the waves. After mating, the female builds a nest and lines it with feathers plucked from her breast. In Iceland, the local people collect the feathers from the nests and use them to make eiderdowns.

MANY KINDS OF CROAKING (pages 16–17)

Some frogs have deep croaks that sound like foghorns, some have high-pitched croaks that sound like birds. Some frogs croak only once and then stay quiet for several minutes, some keep up a constant stream of noise. Croaking is a sort of courtship, because it attracts females to the croaking males.

African Bullfrog. The second largest frog in the world, the male is twice the size of the female. In the dry season, the frogs bury themselves deep in the ground and emerge only when the rains begin. These form the temporary pools in which the frogs breed.

Puddle Frog. The puddle frog comes from the lowland forests of Costa Rica in Central America. These frogs are so-called because they breed in deep puddles formed after torrential rains. The croak of one male encourages other males until the air is filled with a loud chorus which guides the females to one of the puddles.

Painted Reed Frog. During the rainy season in southern Africa, the dips and hollows of the savannah fill with water for a short while. The painted reed frog lays its eggs in these puddles, and the tadpoles must then grow very quickly or be stranded as the water dries up. When the rains start to fall, the males call loud and often, desperate to attract a female and begin breeding as soon as possible.

Red and Blue Arrow-poison Frog. Male arrow-poison frogs croak to lure females to them. Like most frogs they court at night. The males also listen to each other's croaks and fight with any other male who starts making a noise close by. The Indians of South America use the poison from the frogs' skins on the tips of their arrows. That is how the frog got its name.

SNIFFING OUT A MATE (Pages 18–19)

Most mammals do not lay their eggs. Instead the male fertilizes them while they are still inside the female, and when she gives birth some time later, the young are already well developed.

The males cannot see if a female is ready to mate, but when her eggs are ripe, her smell often changes and they can tell that it is time to begin courting.

Klipspringer. A pair of klipspringers define their territory by scent marking. The sight and smell of this dried secretion, produced by both male and female, informs other klipspringers that they are trespassing. Klipspringer, meaning "cliff-hopper," is an Afrikaans word and describes the way in which these small African antelopes spring from rock to rock.

Capybara. Capybaras are the largest animals in the rat family. They live in groups of between ten and forty animals, and in each group there are several males. The males are easy to tell apart from the females by the swollen lumps on their noses. These lumps are glands, and they produce a sticky white perfume that is used to mark out the group's territory. Each male smells slightly different, so the animals in a group can learn about each other's movements just by sniffing.

Dogs. Whenever dogs meet, they begin by sniffing at each other. The smell tells each dog whether the other is male or female, a familiar friend or a stranger. Bitches produce ripe eggs only twice a year, usually in spring and autumn. The smell of a bitch who is ready to mate is very strong.

COURTING DANCES (pages 20–21)

Postman Butterfly. Animals often have more than one way of courting. The male postman butterfly begins his courtship of the female by dancing over her. As he flutters his wings, he sprays a scent over her. His perfume is produced in special pockets on his wings, and smells of flowers. The female then begins to dance too, fluttering her wings. This is the signal to the male that he can land beside her and begin to mate.

Scorpions. The male scorpion holds the claws of the female and dances her backwards and forwards over the same spot. Then he presses a tiny packet of sperm to the ground and dances backwards, pulling his mate toward him. When the sperm packet sticks to her belly, the sperm inside are released to fertilize the eggs.

Wandering Albatross. Young albatrosses are awkward and cannot keep the rhythm of the courtship dance, but with practice they get better and better. After several years of dancing at a nest site, each wandering albatross picks another bird to be its lifelong partner, and they begin to breed. Pairs that have bred together before do not dance as enthusiastically in later years, but get straight to the serious business of raising a chick.

MINIATURE MALES (pages 22–23)

Orchid Mantis. The male orchid mantis spends a lot of his time searching for a mate. He catches less food than the female, who spends all her time hunting. Because the male gets less to eat, and more exercise, he never grows as big as the female.

Panamanian Orb-weaving Spider. The female spider is big and fat because she is carrying masses of eggs. Inside them is all the food for the baby spiderlings. There is no need for the male to grow so big, because the sperm he is carrying are much smaller than the female's eggs.

Garden Spider. Courtship can be dangerous. The instinct of the female garden spider is to attack anything that touches her delicate web. A courting male risks his life as he steps into the trap, for at first the female has no way of knowing that he is anything but a tasty meal. Knowing that he could be eaten by mistake, the male immediately attaches an escape line to the female's web. He then begins to dance along this thread, and the rhythm of his steps tells the female that he is not trapped prey but a suitor come to mate.

COURTING WITHOUT GETTING CAUGHT (pages 24–25)

Glowworm. The glowworm is not really a worm but a beetle. The female just looks a bit like a worm because she has no wings. The light in the female glowworm's tail is made by a chemical reaction beneath the skin. Male glowworms do have wings, and look much more like ordinary beetles. They also have very big eyes, which help them to see the females' strong light.

Mallard. As with most birds, the male mallard is much more brightly colored than the female. The female must be hidden as she sits on the nest, since if she is discovered she might lose her precious young. A male can mate several times in one year, so he dresses in bright colors to impress as many females as possible. He looks much more handsome than she does. But he cannot hide so easily, and if a fox sees them both, the male will probably get eaten first.

WHAT IS IT? (Pages 26–29)

Sage Grouse. The courtship display of the sage grouse is spectacular. Each year, males gather in a favorite spot to show themselves off to the females. Males blow themselves up until they are as big and round as balloons, and then make a deep booming noise that carries for great distances over the open prairies of North America. They fan their tails so that their dark feathers contrast beautifully with their white breasts.

Eventually each female chooses a mate. Every year some males will mate with lots of females, but others may not mate at all. The unsuccessful males must wait another year or two until they have grown big, strong and handsome. Perhaps then they will become fathers.

Great Bower Bird. The great bower bird from Australia builds one of the strangest structures to be found in nature. First it plants an avenue of dead twigs, called a bower. Then it decorates this with a collection of stones, shells, old bones, bits of glass, bottle tops, even seeds, berries and flowers.

The female birds examine the bowers of all the males in their area and mate with the one who has the best display.

PRINTED IN BELGIUM BY proost INTERNATIONAL BOOK PRODUCTION